THEORY OF CREATION
A
SCIENTIFIC TRUTH

CHANDRAN.P

Made with ♥ on the Notion Press Platform
www.notionpress.com

CHANDRAN.P

I was born on 20 January 1966 in Sultan Bathery, Wayanad, to Panthalanickal Krishnankutty and Thankamma. My father was a farmer and my mother was a housewife. Both are no longer alive. My father was renowned in our village for his unwavering honesty, a trait I have inherited and treasure as the greatest wealth he bestowed upon me. I always try to be truthful. When I was a small child, my father told me about a miraculous incident that happened in my father's life because of his prayer to God. Since then, I grew up as a perfect devotee of God.

I pursued a BSc in Physics at St Mary's College, Sulthan Bathery, Wayanad, Kerala. Subsequently, I moved to Chennai and completed AMIE (Electronics and Communication Engineering), a course conducted by the Institution of Engineers. Throughout my academic journey, I was driven by a natural inclination to thoroughly understand and grasp concepts. I chosen the courses according to my interest. I had an innate aptitude for doing technical work.

After completing AMIE, I began my career as an Electronics Lecturer for two years. Subsequently, I transitioned into R&D engineering roles, working with five companies involved in developing and manufacturing medical equipment, automobile parts, and various electronic products. Notable companies I worked with are Schiller Healthcare India Pvt.Ltd in Pondicherry, which develops and manufactures Medical Equipments, and INEL (India Nippon Electricals Ltd), a TVS group company in Hosur, Tamil Nadu, which develops and manufactures Automobile Parts. While at Schiller, the main machines developed are ECG Machine, Treadmill test system, Pulse oximeter and Defibrillator. Worked as AGM R&D at Schiller. Battery Charger, various types of Sensors, Power supplies and Clusters are the main products developed in INEL. At INEL, I worked as Senior Manager, Research & Development.   I retired from INEL at the age of 58.

I possess in-depth knowledge in the subject I am dealing with. Wherever I have worked, I have been able to stand first in the first place. Science and technology are part of my daily life. After becoming a design engineer, I realized that everything in this universe is a design and a design does not exist by itself. It's been more than 20 years since I started saying that the theory of evolution, which is being promoted in the name of science without any scientific evidence, is a nonsense solely based on the similarity between living beings.   A letter written by me criticizing the theory of evolution was published in June 2012 in Sasthragathy magazine published by Kerala Sasthrasahithya Parishad. Over the past 20 years there has been a wide range of reactions, both pros and cons, from those I have discussed. There are those who have asked: what's the point of saying that the theory of evolution, accepted by the scientific world for a century and a half, is wrong. I had an unforgettable experience when a close relative of mine who read what I wrote about evolution and creationism argued with me a lot at first, but later he fully embraced it and he, who was a half-theist, became a full-theist and started living a spiritual life.

I currently reside in Hosur, Tamil Nadu, with my wife and two sons. My wife is a dedicated school teacher. Our eldest son holds a Bachelor's degree in Commerce and is employed as an instructor at a computer center. My younger son is pursuing his degree in Electrical Engineering.

CHANDRAN. P

46/14, 4th Cross, 4th Main,

Manjushree Nagar 2nd Phase,

Near TVS Nagar, Hosur, Tamilnadu- 635 110

Email: chandranpdy@yahoo.co.in

PREFACE

There are two fundamentally different theories regarding the origin of life on Earth – the theory of evolution and the theory of creation. The evolution theory states that all living beings evolved from a common ancestor, and the theory of creation states that God created everything. Darwin's theory of evolution is first understood from the textbook of the 10th class. Later on, learned more about the theory of evolution from the magazines Sasthragathy and Sasthrakeralam of Kerala Sastrasahithya Parishad as well as through online resources. I initially found the evolution theory to be convincing based on a surface-level understanding of its concepts. Now, I am a design engineer with more than 30 years of experience. When I became a design engineer I have come to realize that evolution theory is a flawed concept, while the creation theory is more scientifically grounded. This book explains, through the use of physics, mathematics, and experimental evidence that the God created the entire universe, comprising all living beings and entities, including identical species. The evolution theory relies solely on assumptions based on similarities among species particularly in more advanced species, and it lacks any scientific evidence to support its claims. In fact, the evolution theory is 100% wrong and in contrast, the creation theory, supported by scientific evidence, affirms that God is the universal creator. The word "God" used in this book is a universal term, recognized across cultures and religions, making this book non-religious and inclusive, transcending the boundaries of religion, caste, and creed.

DEDICATION

I lovingly dedicate this book to my parents, Panthalanickal Krishnankutty and Thankamma, who nurtured my faith in God and supported my educational journey, fostering my innate curiosity and passion for knowledge.

CONTENT

1. Introduction…………………………………………...Page 1

2. The Theory of Evolution: A Scientific Blunder………...Page 3

3. Understanding Design………………………………..Page 6

4. Creationism: Scientific Evidence……………………Page 8

4.1 Probability………………………………………...Page 8

4.2 Entropy……………………………………….…..Page 11

4.3 Experimental Evidence………………………….…..Page 12

5. Theory of Evolution: Arguments and Rebuttals………Page 14

6. Creationism: Why It's Often Misunderstood…………..Page 18

7. Creationism: Clarifications and Insights……………..Page 25

8. Theory of Evolution: A Convincing Mechanism……….Page 30

9. Theory of Evolution: Do Minor Variations Lead

    to Major Changes?...Page 35

10. The Parallels Between Computers and Organisms….Page 40

11. Defining God…………………………………….....Page 44

12. The Concept of the Soul………………………….….Page 47

13. Religion and Its Role………………………………..Page 49

14. Ayurveda: A Holistic Approach to Health and Life…….Page 51

15. Conclusion……………………………………….....Page 55

1. INTRODUCTION

For centuries, humanity's most profound questions have revolved around the mystery of existence: Where did we come from, and what is the nature of life? Scientists, philosophers, and theologians have long sought answers, often finding themselves at the intersection of reason and faith.

The origin of life on Earth is explained by two prevailing theories: Theory of evolution and Theory of creation. The theory of evolution states that all living things evolved from a common ancestor, while the theory of creation states that God created everything.

Everyone's belief in God is often rooted in their religious affiliations. There is currently no empirical evidence for God's existence beyond personal faith.

However, this book challenges that notion. It is possible to scientifically prove that God exists and that everything in nature is a manifestation of divine creation. By using mathematics, physics, and experimental evidence, this book provides a novel, empirically driven proof of God's existence and His role in the creation of the entire universe, including all living beings, through simple examples that anyone can understand. No religious concepts are used in this book.

Rather than relying on faith alone, this work bridges the gap between science and spirituality, presenting a perspective that could change how we view the origins of life and the cosmos. It encourages readers to reconsider what is possible when empirical evidence meets profound philosophical questions.

Additionally, the concept of God as a physical entity is explored in Chapter 11.

This book offers insights that could have meaningful implications for the relationship between science and faith, challenging long-held assumptions and inviting readers to explore a new frontier where science and divinity converge.

2. THE THEORY OF EVOLUTION: A SCIENTIFIC BLUNDER

The theory of evolution states that the first single-celled organism appeared in the ocean, and it evolved to give rise to all other organisms. Darwin developed his theory of evolution from the observation that organisms resemble each other. Having observed the similarity between fish, Darwin pointed out that one fish evolved and gave rise to other fishes. Just as, having observed the similarity between a monkey and a human, he said that the monkey evolved and gave rise to the human.

The scientific world is still dealing with the great folly that Darwin said 160 years ago. When fossils are discovered that resemble existing organisms, they are often celebrated as evidence of evolution. In addition to fossil studies, evolutionists rely on molecular biology, genetics, and morphological analysis. Apart from the apparent similarity between organisms, the similarities observed in the internal structure of cells, DNA, RNA, and amino acids among different species are cited as proof of evolution. Despite being one of the largest scientific hoaxes, the theory of evolution remains widely accepted.

The theory of evolution is one of the biggest scientific follies the world has ever seen. In the theory of evolution, only similarity between organisms is presented as evidence for evolutionary processes particularly in more advanced species. The resemblance of one organism to another does not serve as evidence for evolution. A specific pattern and similarity can be found in all of nature's creations (both living and non-living). For example, consider the periodic table of elements, which illustrates the organized relationships and patterns among atoms.

The first atom is the hydrogen atom, composed of one proton and one electron. The next atom is helium, which contains two protons, two electrons, and two neutrons. Thus, the number of protons, electrons, and neutrons in the periodic table follows a specific pattern, increasing incrementally, resulting in the existence of 118 known elements.

After seeing the similarity between atoms, it is a great folly to say that hydrogen evolved into the next atom helium, helium evolved into lithium, and so all other elements emerged through the process of evolution.

If we assume that one species has evolved into another, then the organism of the opposite sex must also have evolved within an observable distance at the same time for reproduction to occur. Otherwise, the species of that creature would not survive. Can evolutionists explain by what kind of evolutionary process organisms of the opposite sex arose in the same place at the same time?

Similarly, there are flowering plants that are pollinated by butterflies. Can evolutionists explain how butterflies and flowering plants evolved in the same place at the same time?

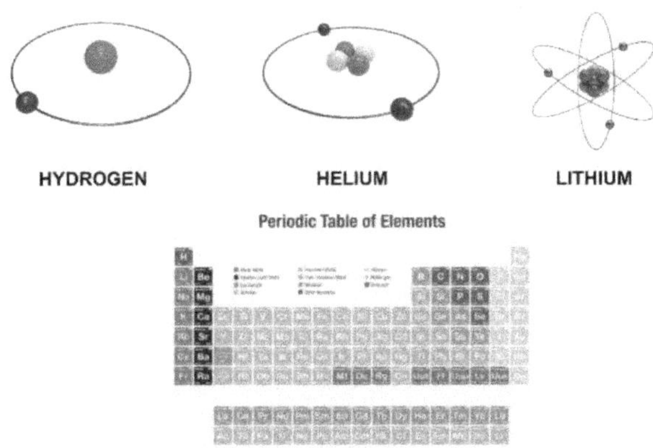

This writer was also once an evolutionist. This author would forcefully oppose creationism, arguing that belief in God should not be invoked to explain the unknown, and instead advocated for the theory of evolution. This writer, who was an evolutionist turned creationist when realized that everything in the universe (including elementary particles, atoms and molecules) is a marvelous design and such intricate design cannot come into existence by itself.

3. UNDERSTANDING DESIGN

Anything that has a certain form or a certain function or both and can be reproduced can be said to be a design.

A straight line can be considered as one of the simplest designs. If twenty or thirty balls are kept in one place, it will never align perfectly in a straight line by themselves.

However, the patterns created by the balls will undergo continuous changes. These spontaneous changes will be uncontrolled changes. The balls will only align in a straight line if a person intentionally arranges them in that way. This principle can be explained using probability theory in mathematics and the concept of entropy in physics, which highlights the natural tendency toward disorder unless directed by an external force.

4. CREATIONISM: SCIENTIFIC EVIDENCE

4.1 Probability

The probability of an event occurring is referred to as the likelihood of that event. Let's consider tossing a coin as an example. At any given time, the coin will either land on its tail side or its head side. Here, the number of favorable outcomes is 1, and the total number of possible outcomes is 2. Dividing the number of favorable outcomes by the total number of possible outcomes gives us the probability of the favorable event.

The probability of getting a tail = the number of favorable outcomes ÷ the total number of outcomes = 1/2 = 50%

Another example is rolling the dice. Dice has 6 faces. Numbers from one to six are written on each face. When the dice is rolled, any number from one to six can appear on top. But only one number will come at a time. Here the number of favorable events is 1 and the total number of events is 6.

So, probability of getting a particular number =

Probability = Number of favorable events ÷ total number of events

= 1/6 = 0.166 = 16.6%

Now let's take the matter of straight line. Let's see what is the probability of 10 balls coming up by themselves in a straight line:

Draw a straight line on the ground and mark 10 points and place 10 balls next to it. A favorable event is when a ball comes to rest at a particular point. The total number of instances the balls can occupy is the total number of points on the ground. The total number of points on the ground is infinite.

Therefore, the probability that a ball will land on a particular point = Probability =1/Infinity = 1/ ∞ = 0

A probability of '0' means it will never happen.

Probability of a ball landing on a point and then another ball landing on another point on the same straight line =

= 1/Infinity x 1/Infinity = 1/ (Infinity)^2 = 0 = 0%.

Probability of 10 balls landing on a straight line

= 1/ (Infinity)^{10} = 0 = 0%

The implication of this is that a straight line will never spontaneously come into existence.

That is, a design, however simple it may be, will never come into being on its own.

Assume that the total number of points on the ground where the balls are placed is finite.

For example, if the total number of points is 1000, then the probability of a ball landing on a particular point = 1/1000.

The probability that the second ball will land next to the first ball = 1/1000 × 1/1000 = $1/1000^2$.

So, the probability of ten balls standing next to each other = $1/(1000)^{10}$ = $1/(10)^{30}$. Understandably, this is "0". (A small approximation is made in this calculation which does not affect the final result).

This means that after a change occurs, the probability of subsequent changes occurring in conjunction with the first change decreases. As the number of changes increases, the probability of it happening becomes increasingly low. Cumulative probability decreases exponentially with each added change.

Anyone who is familiar with the intricacies of life will not find it difficult to understand that the possibility of one living being transforming into another is precisely zero.

According to the theory of evolution, one organism becomes another organism through billions of changes that take place over billions of years. Each of the billions of changes must be consistent with the changes that preceded it. It is not difficult to understand that even a straight line which can be created by just ten changes that are compatible with each other, will not come into being by itself. So, an organism that can be created by billions of changes that are compatible with each other will not come into existence by itself.

Therefore, claiming that small changes will turn into big changes over billions of years and thus different species will come into being goes against the laws of nature and will never happen and is just a plain stupid. Small changes that are incompatible with nature and incompatible with each other will never happen by themselves.

4.2 Entropy

In physics, entropy is the measure of disorder. The second law of thermodynamics states that the total entropy of an isolated system can never decrease over time and can either remain constant or increase. In a system that changes its state on its own, the degree of disorder increases. If the balls are to come in a straight line by themselves, the amount of disorder should decrease by itself. It is not possible to reduce the level of disorder by itself. So, it is impossible for balls to come and stay in a straight line by themselves. The shapes formed by the balls lying on the ground keep changing. Such spontaneous changes are uncontrolled changes.

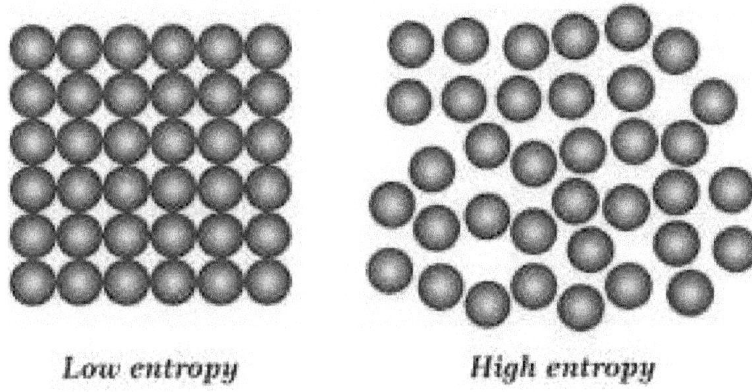

The balls can be arranged in a straight line only if a human gathers them and arranges them.

That is, a design is possible only through the action of an intellect.

Evolutionists say that single-celled organisms appeared in water 3600 million years ago.

Upon closer inspection, what we refer to as a cell is actually a complex combination of various types of molecules organized in a highly specific and intricate manner. In this analogy, a molecule can be compared to a ball; when many types of molecules are arranged in a particular way, they form a living cell. A living cell represents an automatic design that is too complex for human intelligence to fully comprehend. It is important to remember that the single-celled organisms that evolved 3.6 billion years ago contained anywhere from a few hundred thousand to several million molecules.

Even the simplest design—a straight line with 10 balls—cannot spontaneously emerge on its own. Likewise, the most complex design—a living cell made up of millions of molecules, comparable to millions of balls—cannot come into existence by itself.

4.3 Experimental Evidence

Experimental proof involves demonstrating something through experiments and observable results.

All man-made designs serve as examples of such proof. For instance, objects like a pencil, a pen, or a house do not come into existence on their own. They require deliberate effort and design. Similarly, a pencil does not evolve into a pen, nor does a hut transform into a palace on its own. These examples highlight that complex designs require intentional creation rather than spontaneous evolution.

The three commonly used methods in science to prove a concept are mathematical proof, proof using an already established scientific principle, and experimental proof. In Sections 4.1, 4.2, and 4.3, it has been scientifically demonstrated through all these three methods that a design cannot arise by itself. Therefore, it has been proven that a design can only come into existence when intelligence acts.

According to the statement of entropy, the amount of disorder in a system that is undergoing spontaneous change of state will continue to increase. That is, the level of irregularity will continue to increase. If the amount of disorder is to be reduced, i.e. the amount of order is to be increased, an intelligent action is required, as explained in Section 4.2. A design can come about only through the transition from a state of disorder to a state of order. Living things are amazing designs. Therefore, living things can only come into existence through a change from a disordered state to an ordered state. Since the change from a disordered state to an ordered state is impossible to happen by itself, it is impossible for organisms to come into existence by themselves.

Therefore, since everything in this universe is a design and it is not possible for a design to exist by itself, it has been scientifically proven that everything in this nature and universe, including living and non-living, are created by God.

5. THEORY OF EVOLUTION: ARGUMENTS AND REBUTTALS

In class 10th, we learned that the theory of evolution says that the giraffe got its long neck as a result of trying to stretch its head to eat the leaves of tall trees. Those who know the basics of biology can understand that the muscular changes that occur when trying to stretch the neck do not become genetic changes and are not transmitted from generation to generation.

Another argument presented for evolution is that changes occur due to mutations. Mutations are alterations in the genetic structure caused by environmental factors such as ultraviolet rays, X-rays, and chemicals. Experimental evidence shows that mutations primarily lead to biological disorders and health issues, such as cancer; otherwise, they cannot cause evolution. Evolutionists either do not understand these facts or choose to ignore them.

I came across an article by an evolutionist that said that the transformation of tadpoles into four-legged frogs is a repetition of ten million years of biological evolution process from fishes to land quadrupeds. This statement makes no sense. The hatching of the frog's egg and the development of the tadpoles and the subsequent development of the tadpoles into a frog are only different stages in the frog's development. The phenomenon of bodily growth cannot in any way be compared to the concept of evolution.

When a pathogen enters the body of an organism, the immune system activates to destroy it and the resistance against the disease increases. In an evolutionist's article, the increased resistance of bacteria against antibiotics was cited as an example of evolution. However, this is simply a design feature inherent in life, not an example of evolution. Similarly, evolutionists say that the origin of the corona virus is an example of evolution. The cause of the origin of the corona virus is not yet properly understood. Evolutionists cannot understand how ridiculous it is to say evolution is the cause of an unknown event.

Attempts to artificially create life in the laboratory by artificially creating the atmospheric structure that is believed to have caused life in water 3600 million years ago have failed. Even if all the elements necessary for a living cell are present in the air or water, it is not possible for it to organize itself into a particular form which will function in a specific manner (remember what was said about probability and entropy).

Even If all the materials needed to build a house are kept in one place, no matter how many crores of years pass, it will not become a house. Putting all the things you need to make a computer together in one place does not make it a computer by itself. Even a pencil that we use to write will not be emerged automatically. No one has any difficulty in understanding that such human-made designs cannot come into existence by themselves. But scientists and scholars are unable to understand that the "life" which is an extremely complex design and which is beyond the understanding of human intelligence, cannot be emerged automatically. Five important reasons for this are explained in the following chapter.

6. CREATIONISM: WHY IT'S OFTEN MISUNDERSTOOD

(i). From the time we are born, mindset is developed to think that the phenomenon of life is a spontaneous phenomenon. A chicken egg hatches into a chick, which grows and lives. Worms appear to form spontaneously in stagnant water. Plants grow in idle places and life forms there. All living creatures are born, feed, grow, and move about as if life happens entirely on its own. Because of these everyday experiences, people tend to form the belief that life itself is a process that occurs automatically. So, humans are quick to believe that the idea that one organism evolves into another is a very logical, correct idea.

Evolutionists often cite examples like the spontaneous appearance of worms in stagnant water or the growth of plants in empty lands as proof of evolutionary theory. However, these occurrences are not evidence of evolution but rather processes of migration and reproduction. They are design features within the interconnected web of life, not examples of spontaneous or evolutionary development.

Entropy decreases during the condensation of water vapor into liquid water, the formation of ice from water, and the crystallization of minerals from liquids. Evolutionists often present these processes as examples of spontaneous decreases in entropy, arguing that living organisms can emerge automatically without intelligence, even though entropy decreases in these cases. However, these phenomena are integral to the intelligent design of the Earth and its ecosystems, and they should not be interpreted as instances of entropy reduction occurring without guidance.

For example, consider a refrigerator: the entropy inside decreases as the temperature drops. To achieve this, heat energy is pumped out using a compressor, which raises the entropy of the surroundings. Thus, while the entropy within the refrigerator decreases, the total entropy of the system—including the surrounding environment—increases, adhering to the broader principle that entropy tends to rise in closed systems. This reduction of entropy inside the refrigerator is part of its intelligent design. If someone does not understand the heat-removal process and only observes the interior, they might mistakenly believe that entropy decreases automatically. This is similar to how evolutionists claim that entropy decreases spontaneously during the emergence of life, using examples such as water condensation and crystallization without recognizing the influence of an intelligent force behind these processes.

Apart from this, the theory of evolution emphasizes about natural selection. Natural selection states that changes consistent with the environment will survive and thrive. The concept of Natural selection is true. But it is incorrect to claim that small, compatible changes will arise spontaneously and that these minor changes will gradually accumulate into significant transformations over billions of years, ultimately resulting in the development of more advanced living organisms. The truth of natural selection can lead many people, especially those with common sense and knowledge, to believe that small changes will naturally evolve into larger changes over time, especially when presented with examples of natural selection. This notion seems logical at first glance, though it is completely wrong. When this is combined with the idea that 'life is a spontaneous phenomenon,' this belief in evolution becomes firmly established in people's minds.

(ii). If you see a statue of a rabbit made by a small child, it will not be difficult to recognize that it is a statue. Because it will be very different from the real rabbit. But if you see a rabbit statue made by a skilled sculptor, it will look like a real rabbit.

That is, if you see a perfect design, it will be difficult to understand that it is a design. God's creation is so great that humans cannot understand that it is a design.

(iii) Darwin, the originator of the theory of evolution, and scientific scholars have theoretical knowledge only. They have no knowledge of practical science. They have no experience in designing any products. To design something, it is necessary to consider a lot of factors that are interconnected and coordinate them properly. If you want to build a house, first you need to draw a plan, you need doors and windows for the rooms, and you need a washbasin in the kitchen. There should be a tap in the washbasin. There should be a water tank on top of the house to get water from the tap. Water has to be pumped from the well using a motor to supply water to the water tank. In order to have such a design, many things need to be arranged carefully. Creating such a design involves meticulous arrangement and thoughtful planning, demonstrating that even a simple design cannot arise on its own.

 Likewise, a design does not automatically become another design. A pencil does not evolve into a pen by itself, a mobile phone does not evolve into a computer by itself, and a car does not evolve into an airplane by itself.

It is said that if you want to build a house, you must first draw its plan.

That is, before building a house, it has to be decided how the house should be structured.

Each part of the house should be built accordingly. In the plan of the house, where and how each of its parts should be, and what materials should be used for it, everything should be decided in detail. Without doing that, without deciding how the house should look like, if you place the door of the house to one corner and place a window in another corner, dig a well in the middle of the house, fit a motor above the window and build a toilet in the bedroom, it will not become a house.

Similarly, when a creature is created, a clear plan must first be established regarding its final form.

This involves determining how it should live in its environment, how it should move, the structure of its legs, and the design of its mouth, teeth, and eyes. Considerations must also include its dietary needs, how it will escape from predators, and numerous other factors. Each of these elements must be carefully thought out and coordinated to ensure the creature can survive and thrive in its specific habitat.

Just as a house cannot be built without careful pre-determination, the creation of any living being is also impossible without intentional planning and design. Can evolutionists explain how the above inevitable things can happen if a new organism can come into being through billions of years of evolution?

Since evolutionists have no knowledge of what a design is and how it can come about, they only make assumptions that have nothing to do with reality. A significant percentage of people who hear these claims are unfamiliar with the concept of design itself. As a result, they may blindly accept that evolutionists are correct, without questioning the validity of their assertions or considering the complexities involved in the creation of life forms.

(iv). Many people hold the mindset that belief in God is unscientific, which enables evolutionists to claim that any theistic explanation is merely superstition. However, God represents a truth evident to those who seek it. Since science is fundamentally a quest for truth, belief in God is scientific. Rather than being in opposition, theism and science can coexist, each enriching our understanding of existence and the complexities of life.

(v) There may indeed be scientific scholars who recognize the shortcomings of evolutionism. However, they might hesitate to express their views openly due to the fear of being labeled as superstitious or unscientific. This concern about general perception can discourage honest discussions about the relationship between faith and science, leading to a reluctance to challenge established beliefs in the scientific community. As a result, important ideas may not be expressed, hindering a better understanding of these complex topics.

Change is a natural law. Similarly, resistance to change is also a natural law. Isaac Newton's First Law of Motion states that an object will remain unchanged in its state until an external force is applied. This is called Inertia. Inertia also applies to human nature. That is, behavior change requires a deliberate effort.

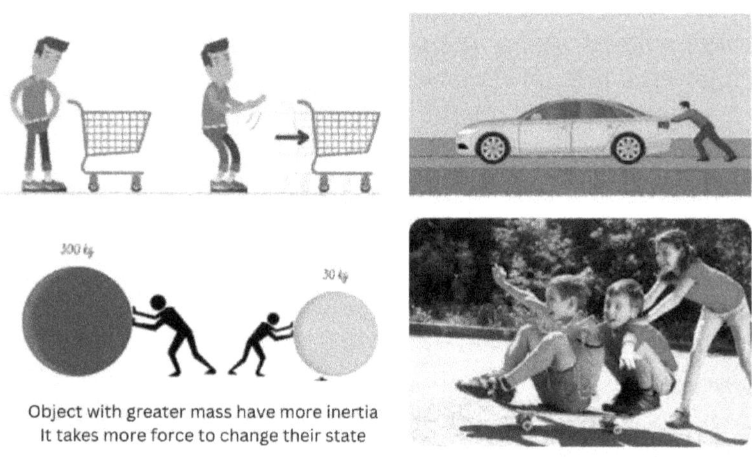

Object with greater mass have more inertia
It takes more force to change their state

Until evolutionists understand what a design is and that a design doesn't come into being by itself, evolutionists will continue to repeat their claims thoughtlessly, like a parrot.

Evolutionists should strive to think realistically, freed from inertia, freed from rigid mindsets. Everyone, including evolutionists, must be able to understand what is meant by design, and understand that the organism is a wonderful automatic design, and that such a design cannot be emerged spontaneously.

7. CREATIONISM: CLARIFICATIONS AND INSIGHTS

It is certain that a wonderful intelligence has worked behind the creation of all living things in this nature. An organism, be it any creature on land, sea or sky, possesses a physical structure and intelligence necessary for its survival. A worm knows how to find and eat its food. A spider instinctively knows how to weave a web, catch its prey, kill it and eat it.

If we examine not only the creation of life but also any creation in this nature and universe (be it plants, animals, air, soil, water, stars etc.) it can be understood that an extraordinary intelligence is working behind it.

For example, let's examine the structure of the atom. Positively charged protons are arranged in the nucleus, while neutrons are cleverly positioned within the nucleus to prevent collapse due to the repulsive forces between protons. Additionally nuclear force is also created inside the nucleus using gluons to keep the nucleus stable. Surrounding this nucleus, electrons revolve in specific orbits. Atoms are designed in such a way that when each atom has 8 electrons in its outermost orbital, the atom becomes stable. So, all atoms try to get 8 electrons in their outer orbital to achieve stability. This fundamental principle drives all chemical reactions, as atoms interact in their quest for this eight-electron structure in their outermost orbit. Consequently, all the basic substances we observe are formed through these chemical reactions.

Atoms and their constituent particles - electrons, protons, and neutrons - have specific roles. Therefore, atoms and the elementary particles within atoms are all wonderful designs. An atom, which serves as the fundamental building block of all matter and functions in a specific way, does not simply come into existence on its own.

In simple terms, the fundamental particles within an atom and the atom itself were created by God. He designed these incredibly small atoms, with even

smaller particles—electrons, protons, and neutrons. Using these tiny building blocks—atoms—God created the entire vast universe, including all living things, by establishing suitable environments and combining them with appropriate natural laws.

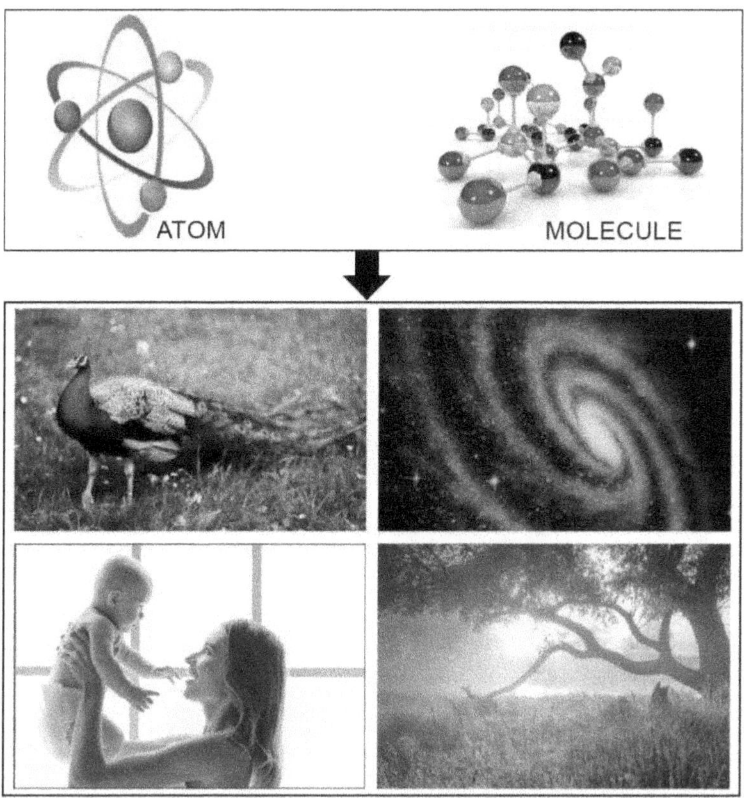

There are 118 types of atoms (92 of which are naturally occurring, while the remaining 26 are artificial) and a certain pattern can be found among all of them, indicating that all atoms share similarities. Why don't evolutionists say that one atom evolved to give rise to other atoms seeing the similarity between atoms? Diversity and unity can be found in all creations of the Creator. For example, there is an analogy between the orbiting of electrons around an atom's nucleus and the orbiting of planets around the sun. Noting this similarity, it is fortunate that evolutionists have not claimed that the solar system evolved from atoms or that the Milky Way galaxy evolved from the solar system.

Evolutionists claim that evolution applies only to living organisms and is based on characteristics such as self-replication, metabolism, and adaptability. In this context, they must explain how single-celled organisms emerged from lifeless molecules around 3.6 billion years ago.

When we speak about the creation of life, it is not the creation of any one particular living being. Rather, it is the creation of a whole ecosystem. When creating a lion, it needs to create the small creatures it needs to eat and other conditions it needs for its survival. Otherwise, the creature will perish. It is not possible to have a completely balanced ecosystem automatically that includes trees, rain, snow, sun and wind and living things on land, sea and sky.

Growth, intelligence and the mind are miraculous phenomena beyond the reach of human intelligence. Think for a moment about the miracle of growth: when an embryo begins developing in the womb, it initially has no organs, no bones, no teeth, and no brain. Yet, as it grows, tissues form, organs develop, bones take shape, and the head and brain emerge. There is absolutely no logic in asserting that the miraculous phenomenon of growth occurred on its own.

Have you ever thought for a moment about the wonder of the brain? It contains approximately 86 billion neurons, forming an intricate interconnected network where communication occurs through electrical and chemical signals. The phenomena of intelligence and the mind are intricately linked to the brain's functioning. The brain of Albert Einstein, one of the greatest intellectuals of the 20th century, is still preserved in the hopes of uncovering the secrets of intelligence. Yet, both intellect and the mind remain enigmatic wonders. Perhaps mind is just another variant of energy fields such as the electric field, the gravitational field, and the magnetic field.

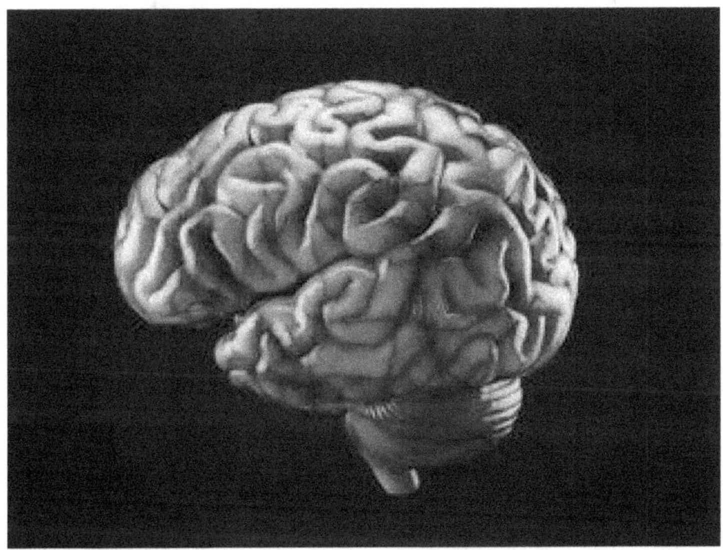

If an electric bulb, 2 wires and a battery are put together in one place, the bulb will not light up by itself by automatically connected to each other by themselves. Can evolutionists explain how the brain comprising an interconnected network of 86 billion neurons, the seat of intelligence and mind and control of so many bodily functions, could have come into existence by itself, when even such a simple electrical network of only four parts could not have arisen by itself? Yet, evolutionists expect us to believe that a system as complex as the brain could develop through the process of evolution over billions of years!

8. THEORY OF EVOLUTION: A CONVINCING MECHANISM

Evolutionists show only mutual similarities between organisms as evidence for evolution, particularly in more advanced species. Mutual resemblance is not a proof of evolution. The theory of evolution lacks experimental evidence and has not been demonstrated through mathematics or any established scientific principles. Evolutionists claim that it is a scientific theory, without the backing of scientific evidence. It is only an assumption. So, it is wrong to call it as a "theory" since it has no support of experimental observations.

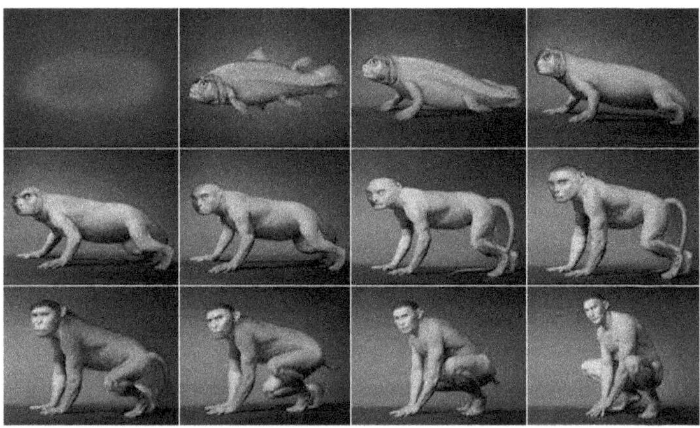

Science is the search for truth. When the theory of evolution was first proposed 160 years ago, it was scientific. Making assumptions to explain events is a valid part of the scientific method, and this is exactly what Darwin did. However, it was a mistake within science—a natural part of the learning process. Unfortunately, the scientists who followed Darwin failed to properly verify and correct the assumptions behind the theory.   Instead of addressing the flaws, they continued to claim that the similarity between organisms was sufficient proof, making misleading but convincing arguments to support the theory. This unscientific approach persists even today, turning the theory into a method of persuasion rather than scientific inquiry. A method of convincing cannot be considered "scientific."

So, what is called as theory of evolution is nothing but an outdated concept, which is 100% unscientific.

Claims such as giraffes got their long necks as a result of trying to stretch their heads to eat leaves on tall trees, transformation of tadpoles into four-legged frog is a repetition of ten million years of biological evolution process from fishes to land quadrupeds,   changes due to mutation cause evolution,   more resistant bacteria and the origin of the corona virus are   due to evolution are all examples of attempts to convince the world without even a trace of scientific backing.

Claiming that 99% of the changes caused by mutation will not persist and 1% will persist is simply a figure mentioned to convince the people easily. Otherwise, this figure makes no sense. Such figures can only be said in the light of statistical study. Can the evolutionists provide explanations for the statistical studies conducted on this subject?

Let me give you a few more ridiculous examples of convincing people to believe in evolution. One article on the theory of evolution says: "In most of the ecosystems available on Earth, there are organisms that can survive in their respective environments. It happened through an evolutionary process called natural selection. This phenomenon cannot be explained except through evolutionary theory. If living beings were created by God, such diversity would not be possible".

He has limited God's intelligence! One can only wonder what kind of device they used to measure God's IQ.

Evolutionists say that "the basis of current knowledge in biology is the concept of evolution and that no knowledge in science has any meaning except in the light of evolution". With this statement, they mean that scientific knowledge can only be meaningful if it is said that "all living things have evolved". If so, all the discoveries made by scientists who do not know about evolution or do not believe in evolution would be meaningless. By this logic, none of these scientists should have conducted research or made breakthroughs.

Evolutionists again say that "evolution is the basis of today's knowledge in modern biology. This is a claim completely lacking in common sense. Biology has progressed not because evolutionists claimed that organisms evolved, but because of rigorous research and discoveries made by scientists over time.

Cancer is a disease caused by abnormal cell division in human organs. Yet, a prominent evolutionist once claimed that cancer is an example of evolution. Such statements are laughably absurd and represent another example of blind persuasion without a shred of logic or scientific validity.

A photo of a one-year-old child and a two-year-old photo of the same child do not show much difference. But there can be a lot of differences between a photo of a one-year-old child and a photo of the same child at 25 years old. This is one of the best examples evolutionists use to convince the world that a small change over billions of years can lead to a big change and the emergence of more advanced species. Since humans of only a few decades of life cannot understand the changes that take place over billions of years through direct observation, many who hear these claims tend to believe them without question. The biggest weapon used by evolutionists is the nonsense of speaking about billions of years. Their ability to use this power full weapon to silence any questioner or critics is second to none.

When you read any explanation of evolution framed in this way, you can see a series of absurdities designed to deceive and convince.

9. THEORY OF EVOLUTION: DO MINOR VARIATIONS LEAD TO MAJOR CHANGES?

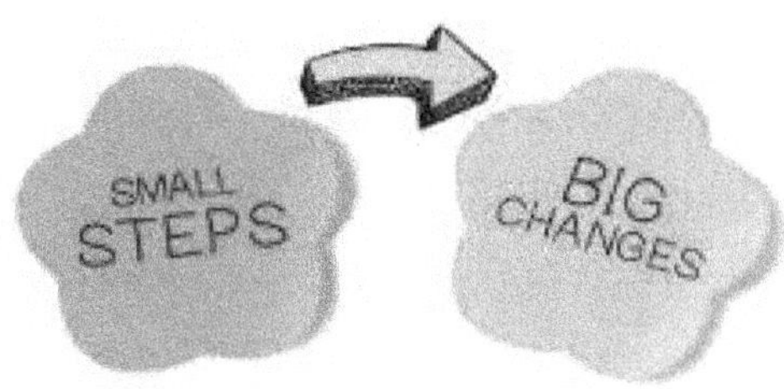

Similarity to one another can be found throughout in nature (God's creation). Similarly, it can be paradoxically said that the uniqueness of each entity in this universe is also a fundamental scientific truth. No two people are alike. No two creatures are alike. It is not possible for even two electrons to be 100% identical. There is a concept in science called "space-time anomaly". It states that "it is not possible for two objects to be in the same place at the same time". That is, no matter how similar two objects are, they are at least different in space-time coordinates.

Evolutionists cite only the similarities between organisms as evidence for evolution.

Similarity of an organism to other organisms, whether it is physical similarity or structural similarity demonstrated by fossil study, molecular biology and genetic science, is not proof of evolution.

Evolutionists are undermining their credibility by claiming that the similarity of one organism to another serves as evidence for evolution. They attempt to fool the world into believing that small changes can accumulate over billions of years to create vastly different kinds of organisms.

It is entirely incorrect to claim that small changes can lead to significant transformations. A small change cannot naturally align with the complexity of nature on its own. For a minor alteration to evolve into a harmonious change, countless complementary adjustments must occur simultaneously in an organized, cohesive manner.

For example, consider the development of legs in fish. Imagine that a bone initially forms. For this bone to develop into a functional leg, numerous additional changes must occur simultaneously. The new bone must connect seamlessly with the body's skeletal system and form the necessary joints. Adequate muscle tissue must also develop to enable movement. Additionally, the brain must undergo changes to produce the electrical signals that activate these muscles along with the development of proper nerve connections to transmit these signals from the brain to the muscles. A functional leg can only emerge when all these intricate changes occur in harmony. Such a complex transformation cannot happen spontaneously or by chance. If a bone forms, and the necessary associated changes take thousands of years, natural selection would not preserve that initial bone over such a vast time span. Furthermore, a solitary bone formed this way would not be inherited by the next generation. For a trait to be passed on, corresponding genetic changes must also occur, requiring millions of molecular-level adjustments. Once a change occurs, the likelihood of subsequent changes happening spontaneously and in a coordinated fashion is implausible (refer to Section 4.1 on probability).

If a living organism is to exist, its design and structure must be predetermined.

Predetermination of structure is impossible within an evolutionary process spanning billions of years (see Section 6(iii) regarding this). For a living being to be purposefully designed, an intelligent force or intellect must be involved.

Therefore, claiming small changes accumulate over billions of years to form significant transformations, resulting in diverse life forms, is completely nonsensical. The scientific world must reject the theory of evolution as nonsense and an outdated myth, realizing that it has no meaning beyond an idea that seems right at first sight because of the similarity between organisms.

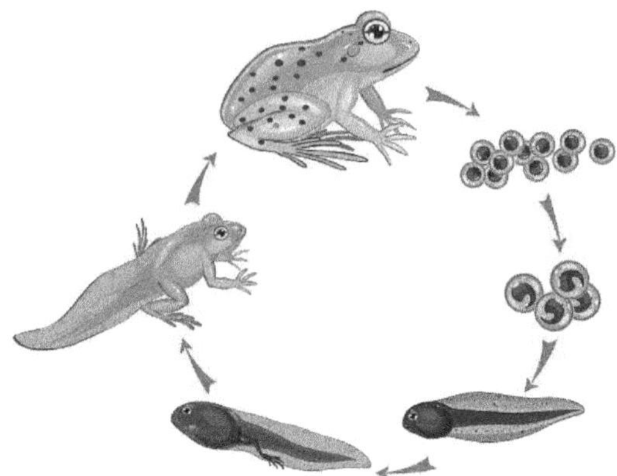

Even if an adaptive change in an animal eventually leads to a genetic change over a long period that results in the development of a different animal, it should still be regarded as a design feature of intelligent design for that specific animal not as an evolutionary process occurring without the involvement of an intelligent force.

For example, consider a self-driving car: it is a vehicle that operates independently, using advanced sensors, cameras, radar, and artificial intelligence (AI) to detect its surroundings, navigate roads, and make real-time decisions for safe driving. These cars are designed to follow traffic rules, detect and avoid obstacles, and safely transport passengers from one location to another. Just as environmental pressures influence change in living things, a self-driving car automatically adapts to road conditions through internal adjustments such as steering control and brake actuation — all enabled by its intelligent design features. Similarly, in nature, environmental pressures can lead to genetic changes if an organism's intelligent design features, such as adaptability as seen in bacteria, allow for such adjustments.

Without these intelligent design features, if a self-driving car's steering and brakes were actuated randomly, it would fail to adapt to road conditions (environmental conditions) and would inevitably crash. In the same way, if random, unguided changes occurred in living organisms without any guided design, it would lead to health problems or even the organism's destruction, and adaptability to the environment would never occur.

It is important to note that there is no empirical evidence supporting the theory of evolution in multicellular, advanced organisms. Thus, the theory of evolution does not apply to More Advanced and Complex Animals. Even in simpler organisms, like bacteria and fruit flies, empirical evidence only supports observable variations within species, not the emergence of fundamentally new, distinct types of organisms. Therefore, the theory of evolution is not applicable to any living organism, whether bacteria, fruit flies, mosquitoes, or more advanced life forms.

10. THE PARALLELS BETWEEN COMPUTERS AND ORGANISMS

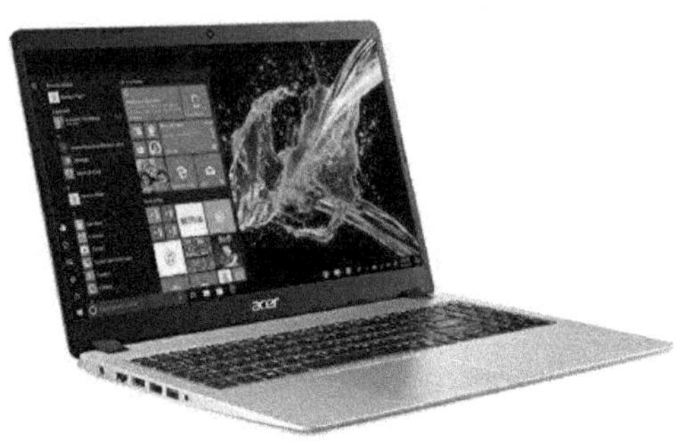

Basically, a computer is made up of components like resistors, capacitors, diodes, transistors and inductors. Ultimately, these parts are all composed of various atoms and molecules. The microcontroller, which is the heart of the computer, contains millions of integrated components of the types mentioned above. None of these parts are automatic by themselves. However, when these non-automatic parts are properly arranged and intelligently connected, they form an automatic device called a computer.

Similarly, the living body is made up of different types of molecules, which, in turn, are made up of different atoms. Different parts of the body are made up of different types of molecules. Atoms and Molecules are not automatic. Just as an automatic device called a computer is formed when non-automatic parts are properly arranged and intelligently connected, an automatic system called a wonderful organism is emerged when numerous non-automatic molecules are intelligently arranged and intelligently connected.

That is, ultimately, everything that we see in this universe, living and non-living, is designed from non-living atoms and molecules. The living and the non-living come into existence according to the difference in design.

Since all living and non-living things are made of non-living atoms and molecules, if mutual similarity is taken as proof of evolution, then the entire universe must have come about through the process of evolution. So then,

can evolutionists explain why all non-living things are excluded from the evolutionary argument and only living things are included.

Note: The chronology of science about when the earth was formed and when life appeared on the earth may be correct. But pointing out that the theory of evolution which states that an organism appeared on the earth by itself and that it evolved by itself to produce all other organisms is wrong. When it says that God created everything, it does not specify when creation occurred, how it happened, whether all living things were created simultaneously or gradually over time, or if different species emerged from modifications of existing ones, etc. If science could uncover the methods of divine creation, it might one day be able to create life itself.

11 DEFINING GOD

Perhaps "God" can be understood as energy. Albert Einstein demonstrated the relationship between matter and energy through his equation $E = mc^2$, showing that matter can be converted into energy. This is what happens in the atom bomb, the nuclear reactor, and the stars. This suggests that matter might have originally emerged from energy. Consequently, the Earth, the universe, and all things within it may have originated from this primal energy. That is everything originated from God. In other words, God created everything from God's own essence. Thus, we can conclude that God is, in essence, the entire universe itself.

We cannot see air or electricity, yet we believe in their existence because we can observe their effects. In the same way, understanding God's presence becomes natural when we recognize that everything in nature, both living and non-living, reflects God's creation. This realization brings complete faith in God. Consider this from another perspective: humans and other creatures possess intelligence, so it follows that an even greater intellectual force must exist within nature and throughout the universe.

The law of conservation of energy states that "Energy cannot be created or destroyed; it can only be transformed from one form to another or transferred between systems." This principle implies that the total amount of energy in the universe remains constant. Therefore, the energy in the universe has always existed and will continue to exist forever. If we equate energy with God, this suggests that God has always existed and will exist forever and God has no birth and death.

Ours is a society where there are many superstitions regarding belief in God. Since the belief in God is scientific, we should not blindly accept that everything associated with it is either true or superstitious. Scientists, religious scholars, and social reformers must take the initiative to clearly differentiate between genuine beliefs and superstitions, striving to uphold the former while eradicating the latter. By doing so, we can foster a prosperous society that embraces faith in God while remaining free from superstitions.

Various Forms of Energy

ENERGY

12. THE CONCEPT OF THE SOUL

Currently, science does not address the concept of the soul. It is true that the soul exists because existence of the soul is affirmed by all religions and all religions are created by God. Instead of categorizing beliefs about the soul as superstitions, science should investigate and strive to understand these concepts through dedicated research. It is important to recognize that labeling the unknown as superstition is not a scientific approach; rather, science is fundamentally about exploring the unknown and seeking truth.

When current enters an electric machine, it starts working. Similarly, when the soul enters the body, the body starts working and the body becomes alive. Therefore, what we refer to as the soul could be a form of energy. Death occurs when the soul leaves the body, just as an electric machine stops when the current is cut off.

Once the cause of the power outage is determined, the cause is resolved and the electrical connection is restored, the stationary electric machine will start working again. Similarly, after solving the physical causes of death, science may be able to revive a deceased person if the source of the departed soul is found and the soul can be brought back.

Religion and Ayurveda are two significant gifts bestowed upon humanity by God at the time of creation. Therefore, these two subjects will be discussed in the following two chapters.

13. RELIGION AND ITS ROLE

When man began to live as a social being, it became necessary to teach him how to live in mutual love and enrich their culture. To facilitate this, God imparted knowledge at birth and created special messengers in many parts of the world, through whom religious texts were written. This is how Christ, Nabi, Valmiki and Vyasa were created in many parts of the world. Through them, texts like Bible, Quran, Ramayana, Mahabharata etc. were written.

The unity and diversity can be found everywhere in God's creation and can also be found in the creation of religion. When religions are considered, unity refers to shared similarities, while diversity highlights the differences among various beliefs. As mentioned in Section 9, the scientific truth that "each is unique" is relevant here. All religions teach humanity how to live, but they differ from each other in the way they teach, in their beliefs, and in their methods of worship practices.

God, who created the infinite universe and living beings from infinitesimally small microscopic atoms, is the all-pervading power of the universe. While praying on such a formless and invisible God, one needs a form or a means to visualize God in their minds. A statue, photo or symbol of God can be used as an aid to help visualize God's form in the mind. These aids can have various names and forms. Any individuals can choose the most suitable representation of the divine power based on their personal faith. Since belief in God is deeply rooted in religious traditions, using these aids in accordance with one's religious beliefs is most fitting for both meditation and prayer.

It should be understood that those who boast that their own religion is supreme, attack other religions and insult their belief systems and worship methods, do not understand the profound mystery of the universal power of creation and are insulting their own religion and the great ideas of their own religion. Human society should properly understand that those who harm other religions are actually harming their own religion and the followers of respective religions itself should carry out appropriate disciplinary actions against such individuals.

Everyone should clearly understand that God is not manifold and that God is only one, though the means of meditation and prayer on God – such as statues, photographs and symbols - have many names and forms.

14. AYURVEDA: A HOLISTIC APPROACH TO HEALTH AND LIFE

God who created humanity also created the plants that he needed for food and medicine. Ayurveda elaborates on the dietary practices and medicinal knowledge essential for maintaining both mental and physical well-being.

During the era when Ayurvedic texts were written, there were no laboratories, and nearly all medicinal formulations derived from various herbs.

From the vast diversity of plant life, it is beyond the capacity of the human mind to systematically identify which plants have medicinal properties, determine the appropriate parts to use, ascertain the necessary quantities, and understand the correct methods of preparation and consumption to treat various ailments.

Therefore, enlightened sages were created by the Creator to share valuable knowledge regarding medicinal plants, the processes for preparing remedies, and the appropriate ways to administer these treatments. Ayurvedic texts were also written by them. Ayurvedic texts like Charaka Samhita, Sushruta Samhita, and Ashtanga Hridayam etc. were written in this way.

At this point, it is important to recognize the distinction between knowledge acquired at birth and knowledge gained through learning. Humans are the only creature that learns and acquires knowledge after birth, uses that knowledge to live and passes it on to the next generations (there are creatures that learn from human training, like police dogs and circus animals, but it should be remembered that that knowledge is not necessary for their survival and it is not passed on to the next generations. A spider weaves a web, a sparrow builds a nest, and a lion can find its prey by using the knowledge that God gives them at birth. Likewise, it is important to understand that the ancient sages who wrote Ayurvedic texts did so through divine knowledge bestowed at birth, rather than through experiments or systematic study.

There are rationalists who call Ayurveda a "pseudo-science" on the grounds that it has not been scientifically evaluated to prove the efficacy of Ayurvedic medicines and how it works in the human body to cure disease. These people are like "donkeys carrying saffron without sensing its fragrance." These are the people who speak science without knowing what science is, without knowing how science should be, without knowing how to scientifically prove something. Ultimately, the quality of any product—whether it is a medicine, an electronic system, a mechanical system, or any form of design—is proven through functional verification. The effectiveness of Ayurvedic medicines, their ability to cure diseases without side effects, has been proven over thousands of years through human use. If their quality has not been demonstrated through experiments and observations and how they cure diseases hasn't been scientifically explained, this reflects a limitation of experimental approaches rather than evidence of ineffectiveness. Everyone must understand this and reject the baseless claims of these so - called rationalists.

Another reason rationalist's attack Ayurveda is because it is based on the concept of the five elements (Pancha Bhuta). Everyone should understand that all the events that happen are true and it is often the explanations for these events that may be incorrect. When the reasoning provided is incorrect or difficult to understand, rationalists tend to claim that the event itself is false. Despite the practical effectiveness of Ayurvedic medicines in curing diseases, rationalists dismiss Ayurveda as false and pseudoscientific simply because they cannot comprehend the Pancha Bhuta concept. This reveals the emptiness of their so-called logic.

When it is said that Ayurveda is based on Pancha Bhuta concept, it should be understood that it is a block level representation and not an explanation at atomic level and Molecular level. When discussing the construction of a house, there is no need to talk about the atoms and molecules involved. It's enough to talk about the different building blocks like bricks, stones, sand, steel, and wood etc. that are used in the construction process.

Science must strive to understand that the five elements—earth, water, fire, air, and sky—do not carry the literal meanings we use in our everyday life and should explore the deeper meanings in which the sages have used these terms to uncover the truths they represent.

Since Ayurvedic medicines are formulated with the knowledge of the divine who knows the body functions completely, and they are based on a comprehensive understanding of the health, Ayurvedic medicines do not have any unhealthy side effects. Another remarkable aspect of Ayurvedic medicine is that, in addition to its ability to effectively cure diseases without side effects, it offers numerous other health benefits for the body. This multifaceted approach is yet another unique feature of Ayurvedic treatments.

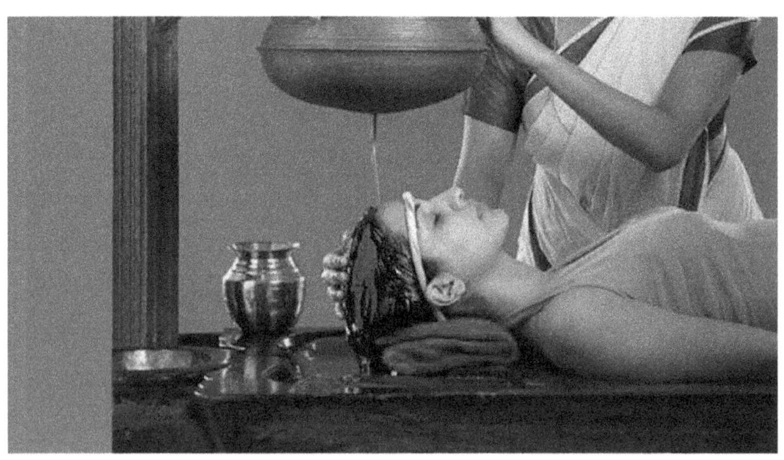

15. CONCLUSION

In the theory of evolution, only similarity between organisms is presented as evidence for evolutionary processes particularly in more advanced species. Similarity to one another can be found anywhere in nature. The resemblance of one organism to another does not serve as evidence for evolution. The assertion that small changes over billions of years will lead to big changes and thus more advanced organisms is just a   statement that seems right at first glance, and it makes no sense and is 100% wrong.

Even the smallest changes, even if it is as minor as a part in a million, will produce meaningful transformations only if they occur in a systematic controlled manner. It is a natural law that orderly changes, regardless of their size, do not occur spontaneously. Spontaneous changes are inherently uncontrolled and cannot produce any form of design, no matter how simple it may be. So, evolution is against the laws of nature and will never occur.

• Everything in the universe, down to the atoms, is part of a purposeful design.

• A design cannot emerge on its own.

• A design can only come into existence through the action of an intellect.

• Ultimately, everything in the universe, both living and non-living, is designed from lifeless atoms and molecules. The distinction between living and non-living forms arises solely from differences in their design.

• Since everything in the universe is a product of design, and design cannot arise spontaneously, it follows that all things, both living and non-living, have been created by God.

• A supreme power exists in this universe. This supreme power is what we refer to as God.

www.ingramcontent.com/pod-product-compliance
Lightning Source LLC
Chambersburg PA
CBHW041647150726
48005CB00015BB/2485